TI-84 Plus CE (

Your Tutor to Learn How the TI 84 CE Works with Screenshots & Keystroke Sequences

Marco Wenisch

For permissions contact:

marco@marcowenisch.com

Did you know that we've also developed an **app** to help you use the **TI-84 Plus CE**? Our app covers most topics from this book and has additional parts like step by step screenshots and video animations of the calculator in use.

www.ti84-calculator.com/manualCE

The app is available for iOS and Android and is called "TI 84 CE Graphing Calculator Manual".

TABLE OF CONTENTS

1 FIRST STEPS

1.1 The Basics of the Basics

When you must press several keys in a row, keys are always pressed one at a time. Never press more than one key simultaneously.

a. Using the 2ND key

Most keys have secondary key functions to access various calculator menus or to input mathematical symbols in an expression. These functions are written in blue and can be used by pressing the 2nd key first.

b. Using the ALPHA key

Above most of the keys, a letter is written in green on the right. Those letters can be used to store a variable or to access shortcuts. Just press the alpha key to access the letters written in green.

To enter many letters consecutively, you can lock into Alpha mode. Then your calculator will only enter the green letters written above the keys. Press 2nd > alpha to activate the Alpha-Lock and press alpha again to exit the Alpha-Lock.

c. Delete and Edit

- To delete parts of your entry, use the arrow keys to place the cursor over the character you want to delete and press del.
- If you want to erase everything you've entered, press the clear key.
- The default mode is to type over existing characters, so just place the cursor over the character you want to replace it with another one.

- The calculator also allows you to insert characters between existing characters. To do so, press 2nd > ins del. Now the calculator will place new characters left to the cursor.

d. Edit Previous Entries

To recall already entered commands or expressions, press the ▲ arrow key to scroll through all previous entries. Highlight the entry you want to use again and press enter. It will be pasted into your current entry line, where you can edit it and do the calculation again.

e. Selecting Menu Commands

Most functions are located inside different menus. They can either have their own button like MATH or STAT, or can be accessed by pressing the 2nd button first. All of the different menus have their own chapter in this book to explain them.

Keep in mind that most menus have more than one page of commands. You can switch between pages with the arrow keys ◄ ►.

To select menu items, one possibility is to highlight them by using the arrow keys ▲ ▼ and hitting enter.

As you can see, the menu items are numbered 1 to 9, and then continued with A-Z. Pressing the appropriate key will also select and paste them into the calculator.

To quit any menu and return to the home menu, press 2nd > quit mode.

1.2 Mode Settings

Choosing the right mode settings is very important in getting the most out of your calculator. In this chapter I will explain all settings in this menu.

If you are in **high school** and do not want to deal with all the different settings, I recommend that you simply select all the settings on the left side. For STAT DIAGNOSTICS only, you would choose ON. Now you have the perfect settings for the most common applications of your TI-84.

Open the mode menu by pressing mode.

Now you can change the settings by highlighting the item in each row. Use the arrow keys to navigate in the mode menu and use enter to select menu items.

MATHPRINT CLASSIC

MATHPRINT mode makes it much easier to work with the calculator as numbers, fractions, and exponents look the same as how you would write them yourself. In CLASSIC mode fractions use a forward slash and look like this: 4/5. Exponents aren't elevated and are displayed this way: 2x^3.

NORMAL SCI ENG:

This setting changes how numbers are displayed on the calculator. For **NORMAL**, numbers are displayed in their usual numeric fashion for up to ten digits. For larger numbers, the calculator will use the scientific mode (**SCI**) automatically. Engineering mode (**ENG**) is just another way to display numbers. I recommend using NORMAL mode.

This is how the calculator will display numbers for the different modes.

(NORMAL: 500000, SCI: 5e5, ENG:500e3)

ENG FLOAT AUTO REAL RADIAN MP

500000
500000
500000
5E5
500000
500E3

FLOAT 0123456789

Here you can set the number of decimal places to be rounded to. For example, selecting **2** will round to two decimal places. **FLOAT** will display as many decimal places as possible.

RADIAN DEGREE

This setting changes how angles are interpreted by the calculator. If you are working with trigonometric functions (such as sine, cosine, and tangent), you would choose **RADIAN**. Then the functions will be graphed for $-2\pi < x < 2\pi$. If you use **DEGREE**, $-360 < x < 360$ would be needed as the limits for the x-axis.

FUNCTION PARAMETRIC POLAR SEQ

These abbreviations stand for Function, Parametric, Polar, and Sequences. Making a selection shows the calculator what type of function to graph.

THICK DOT-THICK THIN DOT-THIN

Choosing THICK means you would like to graph functions with a line, while DOT-THICK will create a graph by plotting only the points, without the connecting line. THIN is just a thinner line.

SEQUENTIAL SIMUL

By selecting SEQUENTIAL, the calculator graphs one function after the other, while it graphs all functions simultaneously in SIMUL mode.

REAL a+bi re^(Oi)

This setting lets you choose between real and complex numbers. For complex numbers, you can use a+bi to display complex numbers in rectangular form and re^Oi to display them in polar form.

FULL HORIZONTAL GRAPH-TABLE

You can choose between different screen modes. FULL is used to display one menu at a time. HORIZONTAL mode splits your display and shows the graph and the home screen at the same time. GRAPH-TABLE displays a graph and a table at the same time. I recommend using FULL all the time as the display is already small enough as it is.

FRACTION TYPE: n/d Un/d

This chooses the fraction type and displays them as simple or mixed fractions.

ANSWERS: AUTO DEC FRAC-APPROX

Using AUTO makes the calculator's answer look like your input. So if you've entered a fraction, it will answer with a fraction. DEC always displays the answers in decimal form and FRAC-APPROX always uses the fraction form.

GO TO 2ND FORMAT GRAPH

This setting only redirects you to the Format Graph screen, which is usually opened by pressing 2nd > format zoom.

STAT DIAGNOSTICS

This will hide or show r and r^2 for regressions in the STAT menu. I recommend turning this ON.

STAT WIZARDS

This setting provides a help window for some commands, like regressions in the STAT menu. It makes it much easier to do the calculations, so you should turn it ON.

SET CLOCK

You can set the date and time by using the arrow keys. However, I wouldn't waste too much time on setting the correct time as you'll probably never need it.

1.3 Basic Arithmetic

When you press on, your calculator turns on and will always show you the home menu where you can do arithmetic operations (if you are in any other menu and turn your calculator off and back on, it will always display the home menu).

Basically, you can enter expressions in your TI-84 Plus like you would write them on paper. However, there are some exceptions that you need to look out for:

a. Negation Key

If you want to enter a negative number, you must use the negation (-) key. It would be wrong to use the subtraction − key and can lead to incorrect calculations or error messages. There's a big difference between the two keys!

b. Enter Exponents

The ^ key with the caret symbol is used to enter exponents. Keep in mind that the cursor jumps up to the exponent position after you press this button and it will remain there until you press the right arrow key. The keyboard of the calculator also includes the x^2 key to square a number.

c. Enter Roots

To enter a square root, press [2nd] > √ [x^2] and type the expression you would like to evaluate. The cursor remains under the radical sign as long as the right arrow [›] key is not pressed.

The symbol for the nth root can be found in the math menu. First, enter the index number of the root and then press [math] > $^x\sqrt{}$ [5] to paste the root symbol to the home menu. Finally, type the rest of your expression.

d. Using Parentheses

You will need to use parentheses for some operations. For example, if you use trigonometric functions like [cos], the calculator inserts **cos(** and you have to type a close parenthesis [)] after the number or expression.

Furthermore, I highly recommend that you always surround negative numbers with parentheses if you want to raise them to a power. Let's say you want to square -2 and enter -2^2 in the calculator; you would expect to see 4 as the answer. However, your calculator will display -4. This is because the negation key [(-)] gets interpreted as -1 * 2 in our example. So the calculator would square 2, which is 4 and then multiply by -1 and the result will be -4. Always surround negative numbers with parentheses.

```
NORMAL FLOAT AUTO REAL RADIAN MP
cos(5)+3
                     3.283662185
cos(5+3)
                    -.1455000338
-2²
                              -4
(-2)²
                               4
```

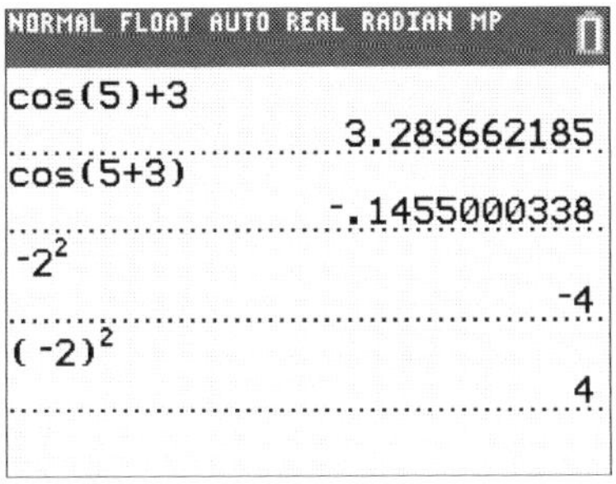

e. Previous Answer

If you start a new entry with an arithmetic operator, +, −, ×, ÷, or ^, the calculator will automatically insert **Ans** which means that the previous answer will be used for the calculation. This won't happen if you start a new entry with numbers or functions. If you want to use a previous answer in your new entry, you can press 2nd > ans (-) to call the previous answer.

1.4 Storing Variables

Storing numbers or expressions can save time if you need them often. For example, if you need to use a number like "1.8634" several times, you can store it as the letter A, so A will be this number.

a. Store Variables

Enter the number or expression you want to store. Then press sto→ which will enter an arrow. Afterwards, enter the letter you want to store it to (press the alpha key and the key belonging to letter you'd like to use).

b. Recall Variables

If you forget what you have stored in a particular letter, you can enter the letter and press enter. The calculator will show its value.

However, if you are entering an expression at the moment and you need to look up a letter without aborting, you can use 2nd > rcl sto→. Then enter the letter you want to look up. Note that the calculator will insert it into your current entry.

1.5 Working with Fractions

As there isn't a fraction key on your calculator, you can only insert real fractions by accessing the fraction function through different submenus. Of course, you can always use the division ÷ key to create a fraction. However, this fraction won't look the way it would when you write it on paper.

a. Enter Fractions

The easiest and fastest way to enter fractions is to open the FRAC shortcut menu by pressing alpha > F1 y=. A menu with four choices will pop up:

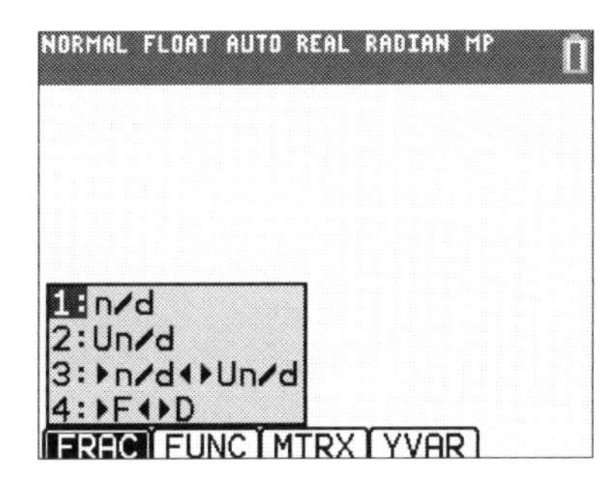

1. **n/d:** Pastes the fraction template to enter common fractions (usually you would use this option).
2. **Un/d:** Just like the first option but to enter mixed numbers and fractions.
3. **>n/d<>Un/d:** Converts a mixed number to a fraction or vice versa.
4. **>F<>D:** Converts decimals to fractions and vice versa.

Use the arrow keys ▲ ▼ to choose between the options and press enter to paste them.

Use the n/d fraction multiple times in the same fraction to enter complex fractions.

$$\frac{\frac{1}{4}}{\frac{5}{2}+\frac{\square}{\square}}$$

b. Converting Fractions

There are two functions to convert a fraction to a decimal (Dec) or to convert a decimal to a fraction (Frac). They are located in the MATH menu, so press math and insert Frac 1 or Dec 2. Confirm with enter.

If you have an infinite repeating decimal and want to turn it into a fraction, you can simply type the first then digits of the number and use the Frac or F<>D command.

1.6 Testing Numbers

The functions of the TEST menu look very simple at first glance, but they can really help you compare complex expressions.

Enter the first expression and press 2nd > TEST math to access the TEST menu. You can choose between six relational operators by pressing the 1 through 6 keys. After that enter your second expression and press enter.

The calculator's answer will be 1 or 0.

- 1 means TRUE
- 0 means FALSE

```
3+2>6
                    0
3+2<6
                    1
```

Your teacher could ask you on a test:

Evaluate ln(2) + ln(4)

1. ln(2-4)
2. ln(2+4)
3. ln(2*4)
4. ln(2/4)

If you don't know the answer, you can use your calculator and test all the answers. The answer 1 means that the statement is true.

```
ln(2)+ln(4)=ln(2+4)
                    0
ln(2)+ln(4)=ln(2*4)
                    1
```

1.7 Converting Angles & DMS

a. Degree to Radian

Make sure your calculator is in Radian mode, and you are on the home screen. First, enter the number of degrees you want to convert to radians. Then press 2nd > ANGLE apps to access the ANGLE menu. Finally, press 1 to paste in the ° symbol and press enter.

If you like to see radian measures expressed as a multiple of π, do the following: Divide the radian measure by π and convert the result to a fraction by using the Frac command math > Frac 1.

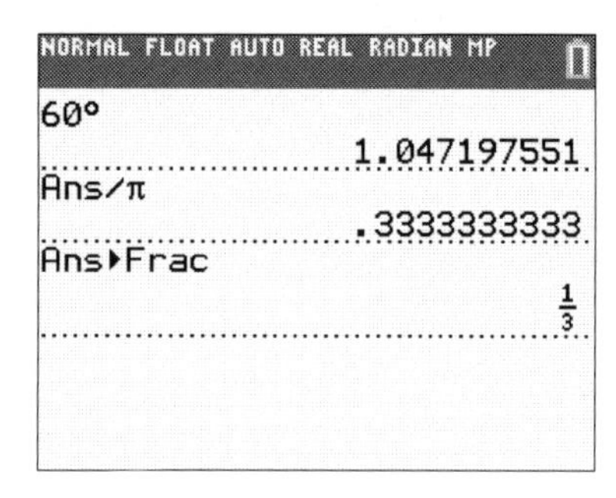

60° in degrees is 1/3π in radians.

b. Radian to Degree

Make sure your calculator is in Degree mode, and you are on the home screen. First, enter the radian measure you want to convert to degrees. Surround the arithmetic expression with parentheses. Then press 2nd > ANGLE apps to access the ANGLE menu. Finally, press 3 to paste in the ^{r} symbol and press enter.

c. Degree to DMS

Make sure your calculator is in Degree mode, and you are on the home screen. First, enter the degree measure. Then press 2nd > ANGLE apps to access the ANGLE menu. Finally, press 4 to paste in the DMS function and press enter.

d. Override Mode of Angles

You can easily force your calculator to use the angle units you want no matter what mode setting is active. To do so enter the number of degrees or radians and add the degree ° or radian r symbol. Press 2nd > ANGLE apps to access the ANGLE menu and press 1 or 3 to paste in the symbol. This will force your calculator to use the units you want regardless of the mode setting:

```
π°▸DMS
                    3°8'29.734"
πʳ▸DMS
                       180°0'0"
```

e. Entering Angles in DMS

Follow these steps to enter an angle in DMS measure:

1. Enter the degree measure and press 2nd > ANGLE apps > 1 to paste in the ° symbol.
2. Enter the minutes and press 2nd > ANGLE apps > 2 to paste in the ' symbol.
3. Enter the seconds and press alpha > " + to paste in the " symbol.

```
30°54'15"
                    30.90416667
```

2 GRAPHING BASICS

2.1 Enter Functions

a. Entering Functions

In order to graph functions, you have to enter them. Press y= to access the Y= editor and enter your first function. After you are finished, press enter which will bring the cursor to the next line. You can enter another function here or press graph to graph the function.

Remember that the calculator only allows the letter X, inserted by the X,T,θ,n key for the independent variable.

You can also use an already stored function when entering new functions. For example, you want to subtract one function from another. Just enter both functions as Y1 and Y2 and enter "Y1-Y2" as function Y3. That makes it clearer and can save time.

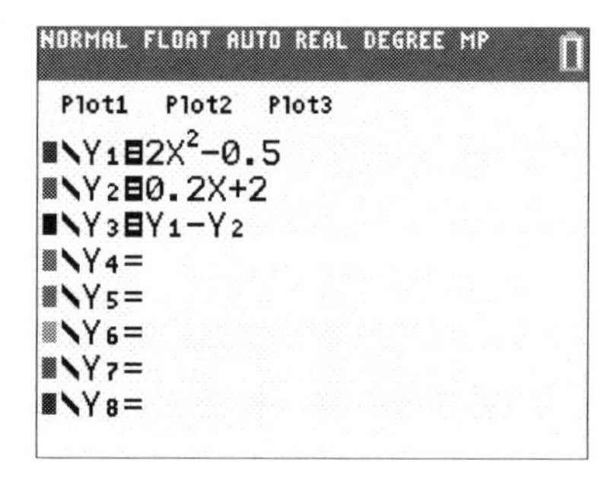

To paste a function name (Y1, Y2, ...), press alpha > F4 trace to open the shortcut menu. Use the arrow keys to select the different functions and press enter to paste them in.

b. Deselect Functions

If you deal with multiple functions, you may want to draw only one function at a time, while the others are still stored. That's absolutely possible and quite easy. To deselect a function, go to y= and use the arrow keys to place the cursor on the equals sign and press enter.

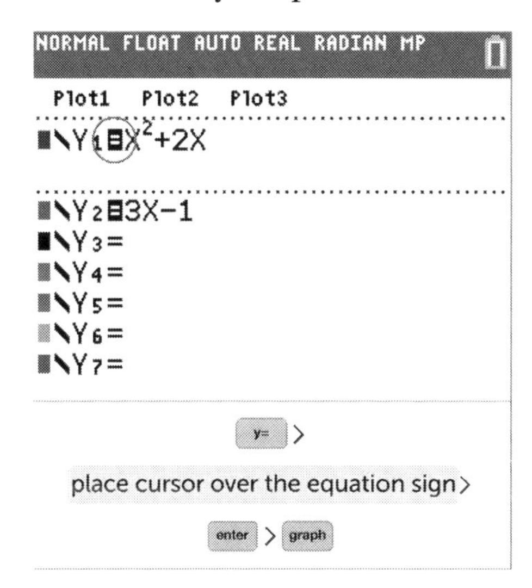

Figure 2-1:

A short equals sign, like you can see for Y1 and Y2 means that the functions will be graphed. A long equals sign, like you can see for Y3, Y4, and Y5 means that those functions won't be graphed.

c. Families of Functions

In case you want to graph a family of functions, enter all numbers for the parameter inside brackets {} separated by commas.

The example below demonstrates the procedure to enter the function f(x)=ax.

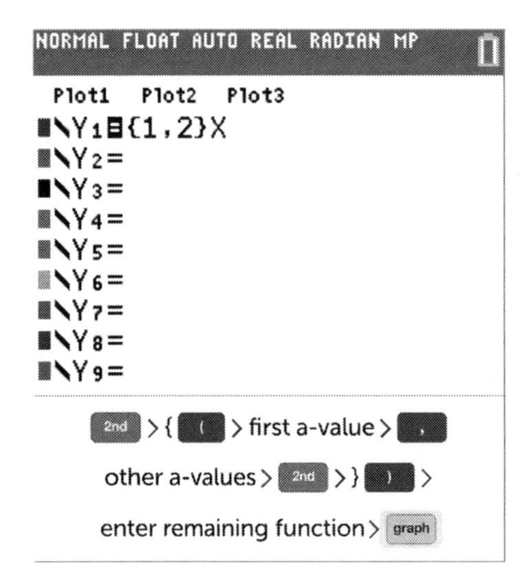

Figure 2-2:

Putting the numbers {1,2} inside brackets means the calculator will draw 1X and 2X.

2.2 Formatting the Graph

The settings made in the FORMAT menu will affect the appearance of the GRAPH menu.

Open the FORMAT menu by pressing 2nd > FORMAT zoom.

Now you can change the settings by highlighting an item in each row. Use the arrow keys to navigate in the mode menu and use enter to select menu items.

The best way to go is to select all the options on the left. If you want to know more, read the following section, which explains all of the format settings.

RectGC PolarGC

You can tell the calculator to display the coordinates of the cursor in rectangular (**RectGC**) or polar form (**PolarGC**). Unless you work with complex numbers, select RectGC.

CoordOn CoordOff

ON means the calculator will show the coordinates of the cursor at the bottom of the screen. If you turn this off, you won't see the coordinates anymore if you trace a function.

GridOff GridDot GridLine

Choosing GridDot adds a grid to the GRAPH menu, which you probably won't need. GridLine adds a grid in form of a line. I find them more distracting than useful and recommend to select GridOff.

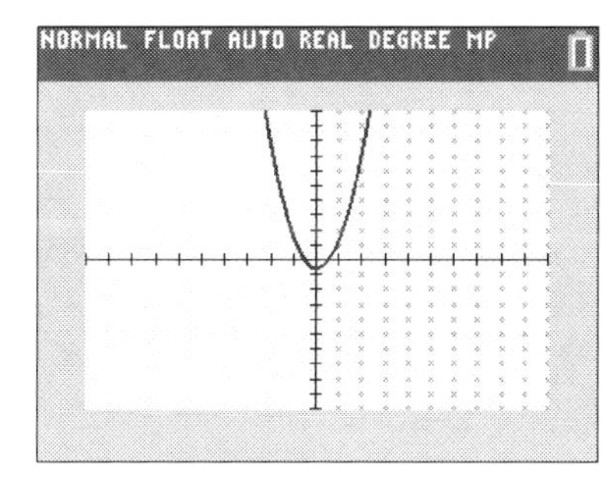

GridColor

Here you can choose a color for the grid if you've set GridDot or GridLine in the setting above.

Axes

Ability to set the color for the x- and y-axis.

LabelOff LabelOn

You can label the axes with x and y by turning LabelOn. However, it doesn't make much sense as the location of the labels isn't very good.

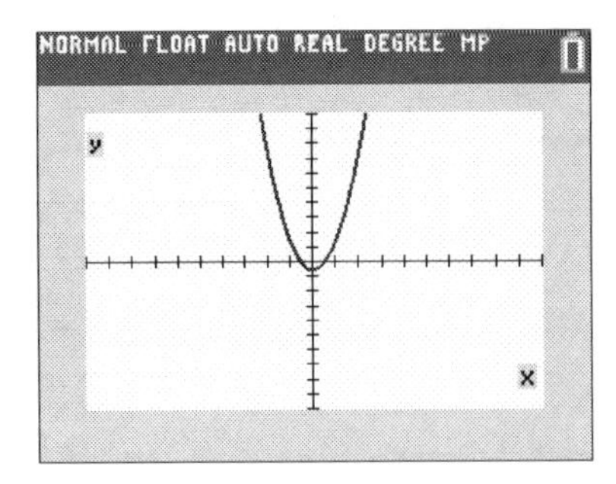

ExprOn ExprOff

This setting is effective if you trace functions (more than one). It will show the definition of the functions in the upper-left corner (ON) or hide it (OFF).

BorderColor

You can change the color of the border (the area beside the coordinates and the graph).

Background

You can choose an image which will be used as a background or select a color.

Detect Asymptotes On Off

By selecting On the calculator won't draw vertical asymptotes. If you select Off, the calculator connects all points of the graph while asymptotes are not connected in real.

2.3 Setting the Graph Window

Your calculator is limited to the x- and y- coordinates set in the WINDOW menu when it graphs a function. Therefore, you should know how to properly set the data in this menu.

Open the WINDOW menu by pressing window.

Use the arrow keys to navigate and simply override existing values by entering new numbers, or press clear and enter a new number.

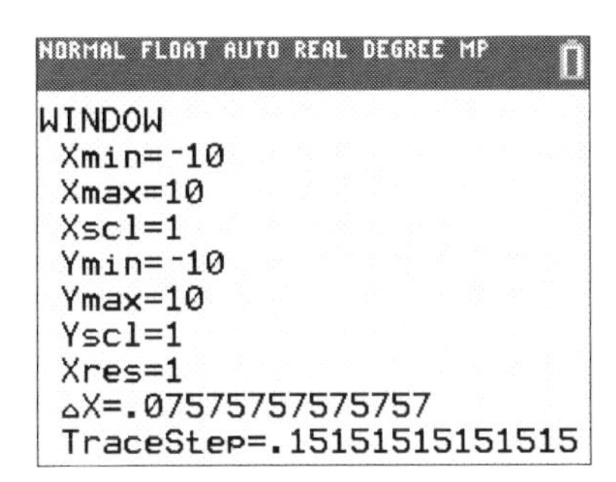

Xmin & Xmax

Sets the start and end value of the x-axis. Make sure Xmin is smaller than Xmax.

Xscl

You can change the distance between the tick marks located on the x-axis. The smaller the number, the smaller the distance between them and the more tick marks will appear.

Ymin & Ymax

Sets the start and end value of the y-axis. Make sure Ymin is smaller than Ymax.

Yscl

You can change the distance between the tick marks located on the y-axis. Remember that the calculator uses your entered numbers in absolute value, so it doesn't make a difference whether you enter 1 or -1.

Xres

This option accepts integers from 1 to 8 and determines the resolution of the graph. 1 means the highest resolution and takes the most time for the calculator to plot the graph. Usually set Xres to 1 and only increase the number if the calculator takes a long time to graph your functions. By increasing the number, the graph becomes less accurate.

ΔX

This variable determines the size of the steps when tracing the x-axis. It sets the amount the x-value changes each time you move the cursor right or left.

TraceStep

The TraceStep is connected to the value for ΔX and is always twice the amount of it.

2.4 Zooming the Graph Window

Using the ZOOM functions of your calculator makes it easier to adjust your graph window than setting the coordinates in the WINDOW menu. You can choose between several commands which will be explained in this chapter.

Open the ZOOM menu by pressing zoom. Notice that you don't have to graph your functions before using ZOOM commands. You can access the ZOOM menu directly from the Y= menu, just select a command and it will subsequently be graphed.

a. Zooming the Graph

ZBox

Allows you to construct a box which restricts the coordinates of the new window. Press zoom > 1 and use the arrow keys to move the cursor to the first corner of the box. Press enter to set the corner and do the same again to set the second corner. Finally, press enter to confirm the box or press clear to start over.

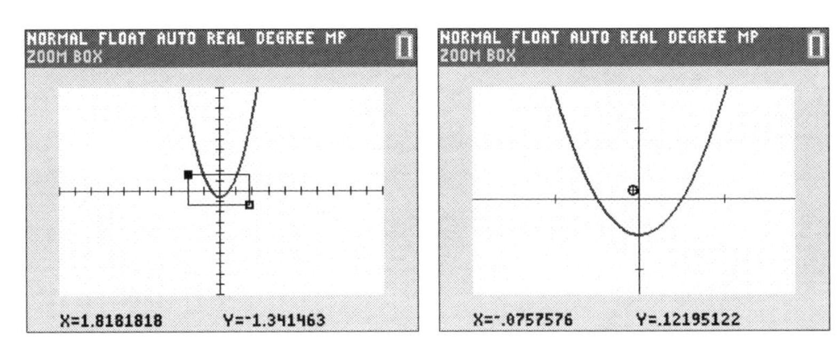

Zoom In & Zoom Out

Zoom in and out from the graph by pressing enter. It will zoom based on the location of the cursor. Move the cursor by using the arrow keys to zoom to a different location.

ZDecimal

This command zooms the window to the following coordinates: $-4.7 \leq x \leq 4.7$ and $-3.1 \leq y \leq 3.1$.

ZSquare

Your current window will be readjusted to graph elements like a circle properly. Setting the x and y coordinates to the same value would distort the circle as the calculator screen has a larger width than height.

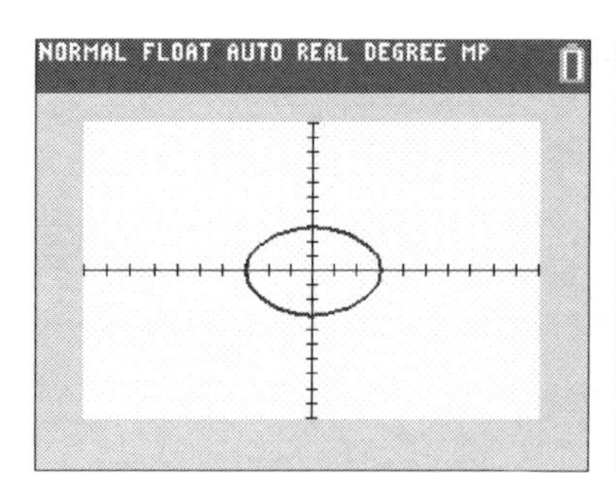

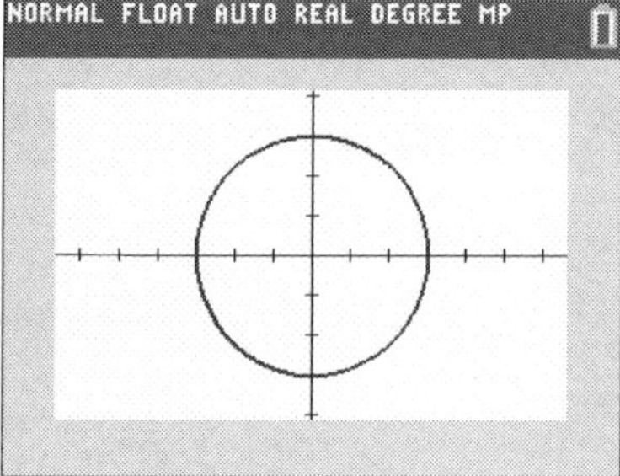

To draw a circle, start from the main menu and press 2nd > DRAW prgm to access the DRAW menu. Select Circle 9 to paste the command to the main menu. Then enter your X- and Y-value, separated by a comma and specify the radius of the circle. Press enter to draw the circle.

For example: Circle(0,0,3). This will draw a circle with a radius of 3 and a center point of (0|0).

ZStandard

This command zooms the window to the following coordinates: $-10 \leq x \leq 10$ and $-10 \leq y \leq 10$. If you don't know the dimensions of your function, ZStandard is always a good option to start with and, when needed, you can use ZoomIn or ZoomOut.

ZTrig

Use this command to set the window when graphing trigonometric functions. It provides the best settings for such functions, like setting the tick marks on the x-axis to $\pi/2$.

ZoomStat

Only use this command to find a suitable viewing window if you are graphing plots. For normal functions, it will be useless.

ZoomFit

This command automatically zooms the y-coordinates of the window to fit your graph well. It doesn't affect the x-axis.

ZQuadrant1

This command zooms the window to the following coordinates: $0 \leq x \leq 9.4$ and $0 \leq y \leq 9.4$. Consequently, it will only show the part of your functions in Quadrant I.

b. Stop and Undo Zoom Commands

If you use a ZOOM command and you see that you won't like it while it is being plotted, you can abort it by pressing [on]. This will save you the time it would take for your calculator to finish plotting.

You can undo a zoom command by accessing the ZOOM MEMORY menu. Press [zoom] > MEMORY [▸] and select ZPrevious [1].

c. Storing a Zoom

You can store and recall your own favorite graphing window, which will remain on your calculator even if you turn it off. Note that you can only store one window setting at a time.

Press [zoom] > MEMORY [▸] and select ZoomSto [2] to store your current graphing window settings.

To recall graphing window settings, press [zoom] > MEMORY [▸] and select ZoomRcl [3].

3 DIFFERENTIAL CALCULUS/ ANALYZING FUNCTIONS

3.1 Tracing a Graph

After you have graphed your functions, there are two different ways of tracing them.

1. Free trace by using any of the arrow keys. This method will show you the x- and y-coordinates of any point located within your graphing window. However, those points don't have to be an actual point on the graph.
2. The second tracing method is to press trace and only use the right or left arrow keys to investigate the function. Most of the time you will use this method.

While you are tracing the graph, you can enter any x-value and the cursor will jump to that point. Make sure the value is between your window settings for Xmin and Xmax (so that you can see the point in your graphing window).

You can change the amount the x-value changes each time you move the cursor right or left. Press window and move your cursor all the way down to the ΔX line. Override the value with your desired trace step value.

3.2 Find Y-Value

Press 2nd > CALC trace to access the CALC menu. Select value 1 and enter the x-value of the point where you want to find the y-value.

Make sure the display of the calculator is showing your entered x-value. Otherwise, you will get the error: "INVALID".

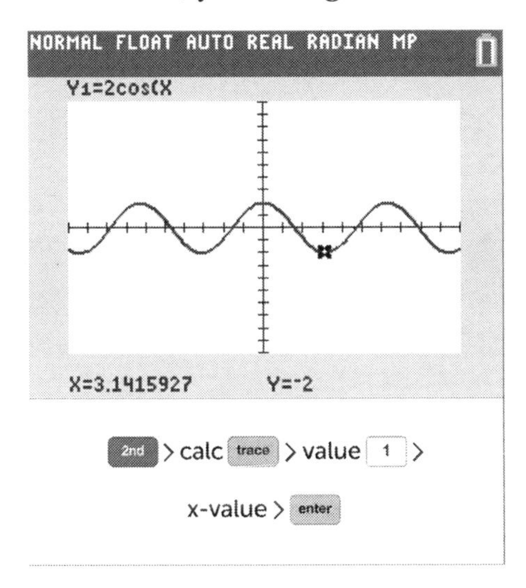

Figure 3-1:
The screenshot shows the y-value of the function Y1=2cos(X) at the point X=π.

3.3 Find X-Value

There is no function to directly enter a y-value and find the x-value of the function at that point. You have to enter your y-value as a new function and find the intersection of both functions. This will show you the x-value.

Press 2nd > CALC trace to access the CALC menu. Press intersect 5, select both functions and make a guess by pressing enter.

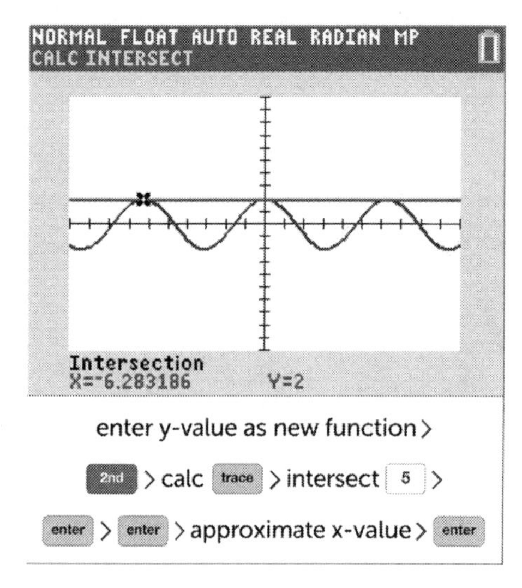

Figure 3-2:
Finding the x-value of the function Y1=2cos(X) at the point Y=2.

3.4 Y-Intercept

A quick and easy way to find the intersection with the y-axis (x=0) is to paste the name of your function into the home menu. Go to the Y= editor and enter your function, then follow the steps below to find the y-intercept.

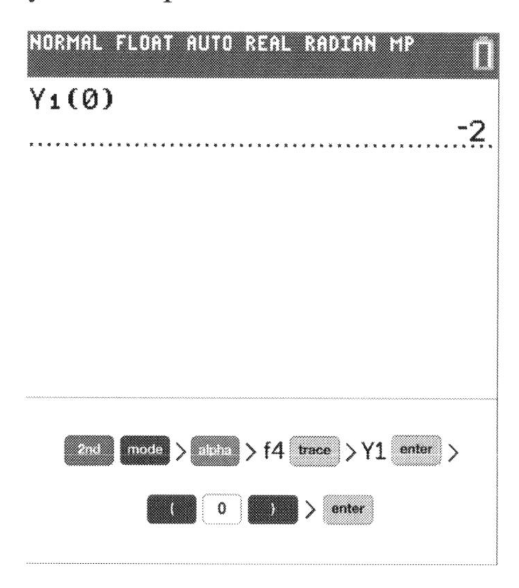

Figure 3-3:
Finding the y-intercept of $Y1=0.5X^2-2$.

3.5 Zeros of a Function

The zeros of the functions are the points where the graph of the function intersects or touches the x-axis. In addition, they are the solutions of the equation f(x)=0.

Press 2nd > CALC trace to access the CALC menu. Select zero 2 and set the left bound by using the arrow keys to place the cursor left of the zero. Alternatively, you can enter an x-value and press enter. Do the same for the right bound. Finally, guess the approximate position of the zero and press enter.

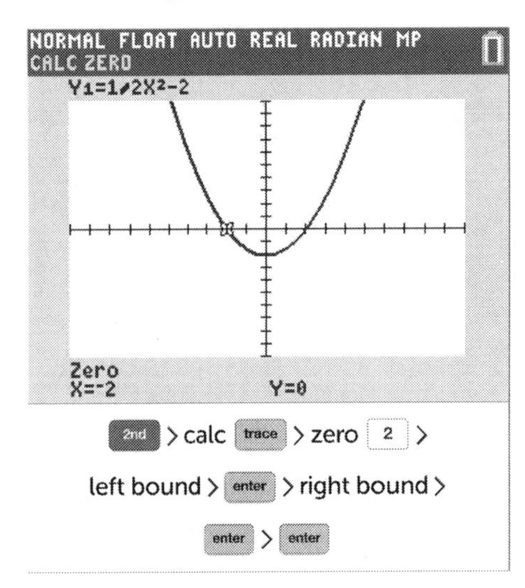

Figure 3-4:

One zero of the function $Y1=0.5X^2-2$.

3.6 Minimum

Press 2nd > CALC trace to access the CALC menu. Select minimum 3 and set the left bound by using the arrow keys to place the cursor left of the minimum. Alternatively, you can enter an x-value and press enter. Do the same for the right bound. Finally, guess the approximate position of the minimum and press enter.

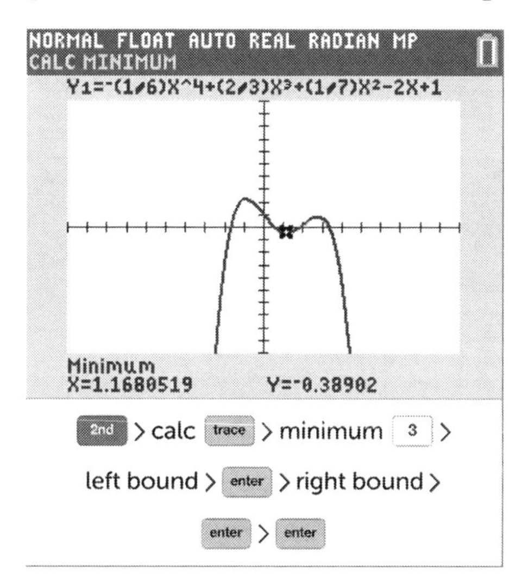

Figure 3-5:

Finding a minimum (turning point).

3.7 Maximum

The way to find a maximum is exactly the same as for a minimum with one difference: to select maximum press 4 in the CALC menu.

Press 2nd > CALC trace to access the CALC menu. Select maximum 4 and set the left bound by using the arrow keys to place the cursor left of the maximum. Alternatively, you can enter an x-value and press enter. Do the same for the right bound. Finally, guess the approximate position of the maximum and press enter.

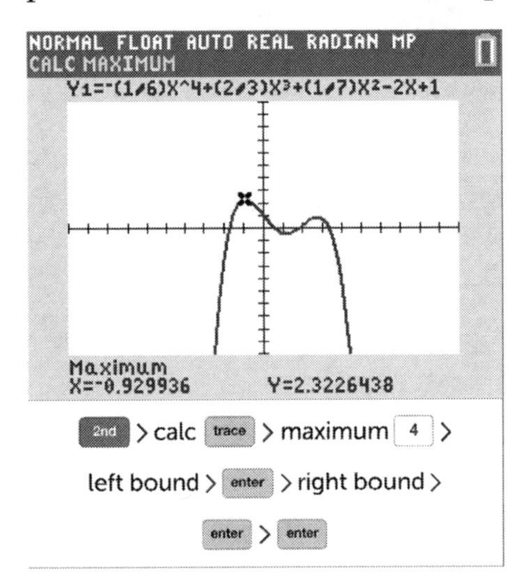

Figure 3-6:
Finding a maximum (turning point).

3.8 Intersection of two Functions

Press 2nd > CALC trace to access the CALC menu and select intersect 5. Now the first function must be selected. Press enter to select it or use the up and down arrow keys to scroll through the functions. Do the same for the second function (if you only have stored two functions, you can press enter two times). Finally, make a guess near the point of intersection and press enter.

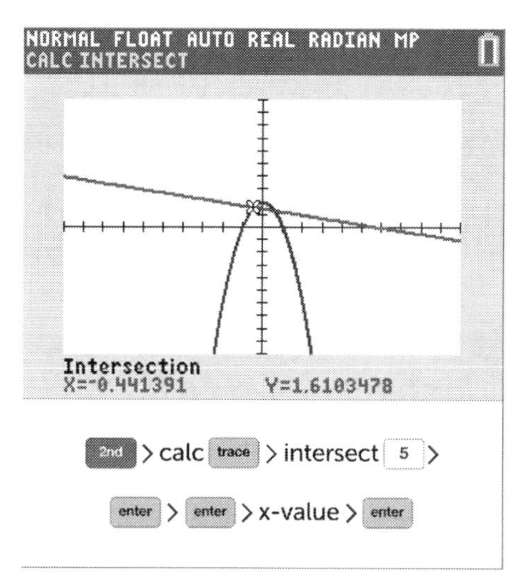

Figure 3-7:

The intersection of the two functions Y1=-2X²+2 and Y2=-0.25X+1.5. Note that there is a second intersection right of the y-axis.

3.9 Draw Derivation

You can draw the derivative by entering it as a new function in the Y= menu. Press math > nDeriv 8 to paste in the template and press X,T,θ,n to set the variable, which will be "x". Subsequently, press alpha > F4 trace and choose your function. Press X,T,θ,n once more and press graph to draw the derivative. Use the arrow keys to navigate through the template while entering the function.

If you only want to know the slope at one point of the function, press 2nd > CALC trace to access the CALC menu. Select dy/dx 6 and use the arrow keys to move the cursor to your desired point and press enter. Alternatively, you can enter an x-value and press enter.

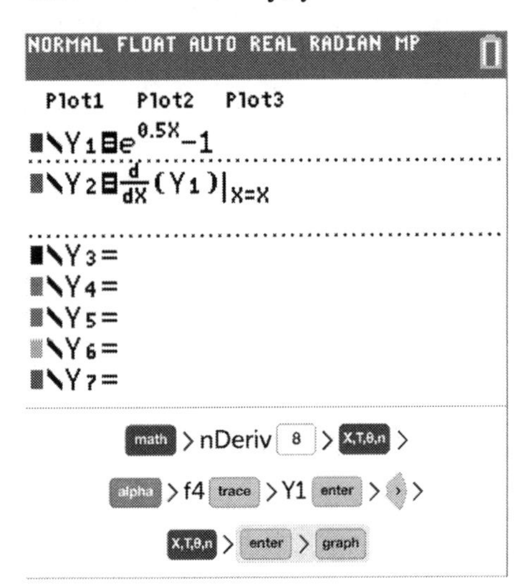

Figure 3-8:

Syntax to enter the derivative as a new function Y2.

3.10 Inflection Points

The condition for inflection points is f''(x)=0. The turning points of the first derivative show the zero of the second derivative. Thus, you can determine the x-value of the inflection point by searching for the maxima or minima of f'(x). If you look at the graph of the function, you can decide whether it's a saddle point or an inflection point.

What you have to do step by step:

1. Enter the derivative of Y1 as a new function.
2. Find all turning points (minimum or maximum) of the derivative and note down the x-values of those points.
3. Find the y-value of that x-value of Y1. This will be the inflection point.

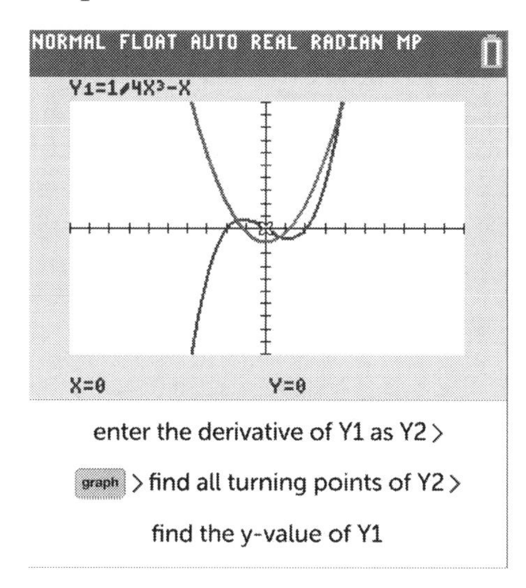

Figure 3-9:
The inflection point of the function $Y1=0.25X^3-X$, which you can easily find by looking for the turning points of the derivative of Y1.

3.11 Calculate Tangent and Draw Line

Press graph to graph your function as this command won't work from the Y= menu. Then press 2nd > DRAW prgm to access the DRAW menu and select Tangent 5. Enter the x-value where the tangent touches the function and press enter.

The equation of the tangent line is shown under the x-coordinate in the last picture.

Keep in mind that there are additional commands in the DRAW menu. However, most of them are not very useful. Feel free to try them out to see if you can use them in your calculations.

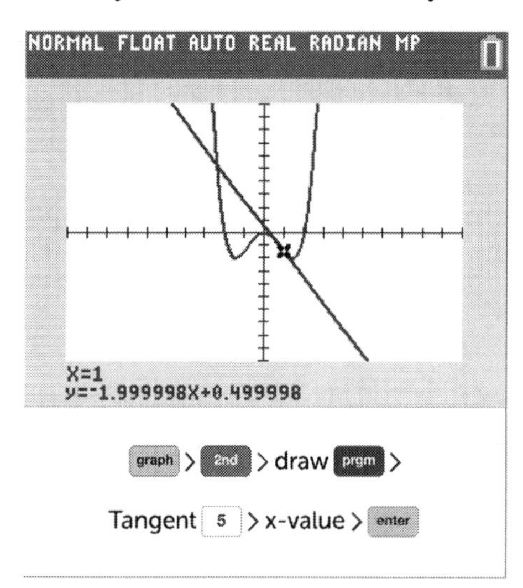

Figure 3-10:
Tangent line of the function at x=1, showing the equation of the tangent.

4 SOLVE EQUATIONS

4.1 Polynomial

Insert your polynomial equation as a function in the y= menu. Make sure the polynomial equation equals 0. Press 2nd > CALC trace to access the CALC menu. Select zero 2 and find all zeros of the function, which are the solutions to your equation.

Read the chapter Zeros of a Function for additional information.

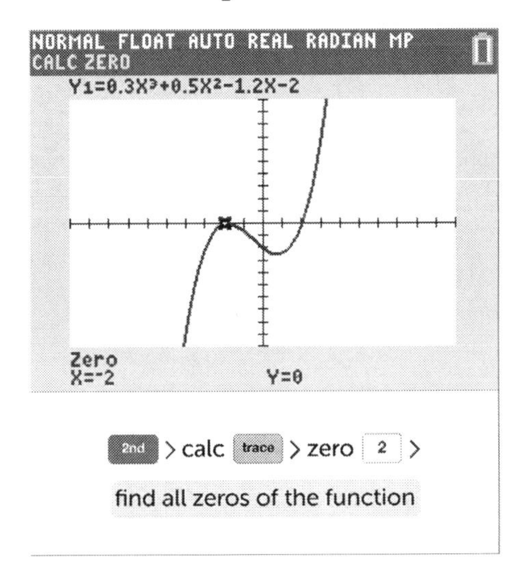

Figure 4-1:

First zero (solution) of the graphed polynomial equation $0.3X^3+0.5X^2-1.2X-2=0$

4.2 Solve any Equation

You can graphically solve any equation by doing the following: Access the y= menu and enter the part of the equation as Y1 which is left of the equal sign. Enter the right part of your equation as Y2. Press 2nd > CALC trace to access the CALC menu and select intersect 5. Then find all points of intersection, which are the solution for the equation.

Go to the chapter Intersection of two Functions to read more about how to find the intersection of two functions.

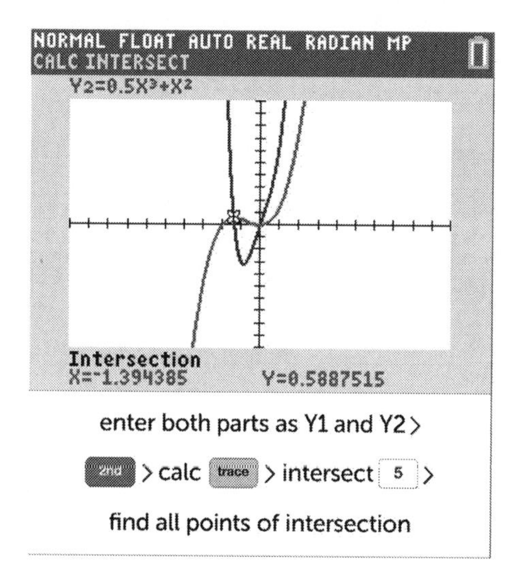

Figure 4-2:
First intersection (solution) of a graphed equation.

4.3 Equation Solver

The Equation Solver is a tool that can be used to solve equations with one variable. If your equation has more than one solution, it is a bit tricky as you have to play with the guess and bounds. I would use the way described in the previous chapter instead to find solutions of any equation quickly.

a. Transform and Enter

To access the Equation Solver, press math > ▲ > enter. You will see two empty boxes, E1 and E2. If they are not empty, press the button clear first.

Let's take 2(X+1) = (6X+0.5) as an example. Simply enter the first part of the equation (left of the equals sign) in the box "E1" and the second part in "E2".

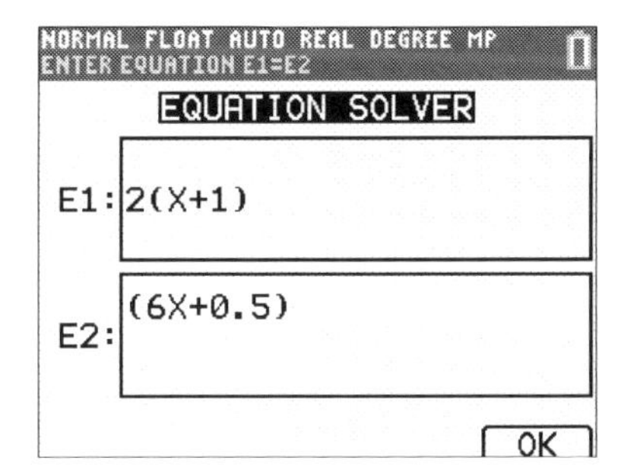

b. Solve Equation

After you have entered the equation and pressed enter or OK graph to confirm it, you must guess the solution. Starting from this value, the calculator begins to solve the equation. If your guess is close to the solution, you will get the result faster. Guess 1 for the line "X=" if you don't know what to enter. Confirm your guess with enter.

The next step will be to set the bounds. For equations with only one solution, you can use the default setting {-1E99,1E99}. Your calculator should look like this:

```
2(X+1)=(6X+0.5)

  X=1
  bound=█-1E99,1E99}
  E1-E2=0
```

To solve the equation, use the arrow keys ▲ ▼ to place the cursor in the line of the variable X. Then press alpha > SOLVE enter. This will override your guess for X with the calculated solution.

```
2(X+1)=(6X+0.5)

▪X=.375
  bound={-1E99,1E99}
▪E1-E2=0
```

After you have calculated X, it can be used on the home screen to do further calculations. One possibility would be to check if the solution can be converted to a fraction.

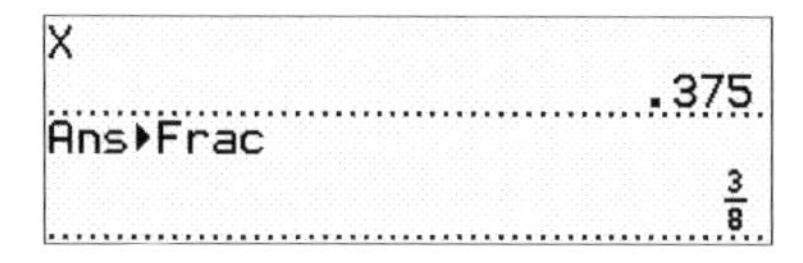

c. Find Multiple Solutions

Functions with higher degrees especially tend to have more than one solution. There are two ways to find all solutions with the Equation Solver.

1. Guess a large positive or negative number. For example, if the equation has two solutions, 1 and 5, and you guess 10, it will find the '5' solution. If you guess -10, it will find the '1' solution (it will always find the solution that is closer to the guessed value).

2. Redefine the bounds after you find a solution. After you find the '5' solution and you are sure it is the largest solution, set the upper bound to 4.99. This makes sure that the calculator searches for other solutions smaller than 5. Also, redefining the bounds for trigonometric functions can be useful to find solutions within a certain interval.

If your guess is not within the bounds, you will get a BAD GUESS error. Change the value of the guess and check the bounds to fix it.

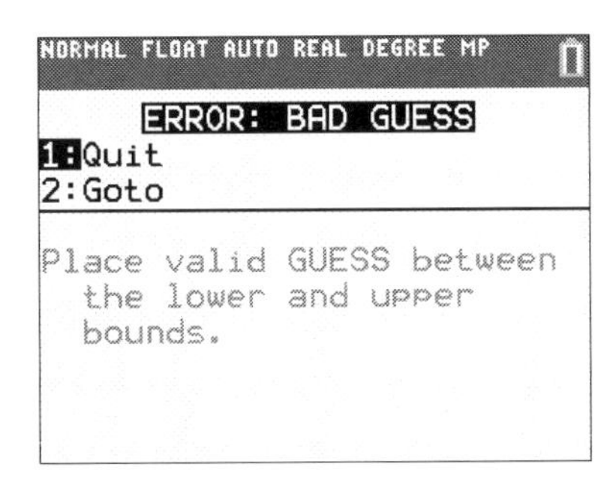

5 INTEGRAL CALCULUS

5.1 Calculate Integral

Start from the home menu and press math > fnInt 9 to paste in the integral template. Enter the lower limit and use the right arrow key to move the cursor to the upper limit. Press again and enter your function. Finally, use to move the cursor behind the "d" where you have to press X,T,θ,n to set the variable which will be "X". When you are done, press enter to calculate the integral.

Instead of entering a new function you can also insert "Y1" to get a function you've already stored. Press alpha > F4 trace and choose your function.

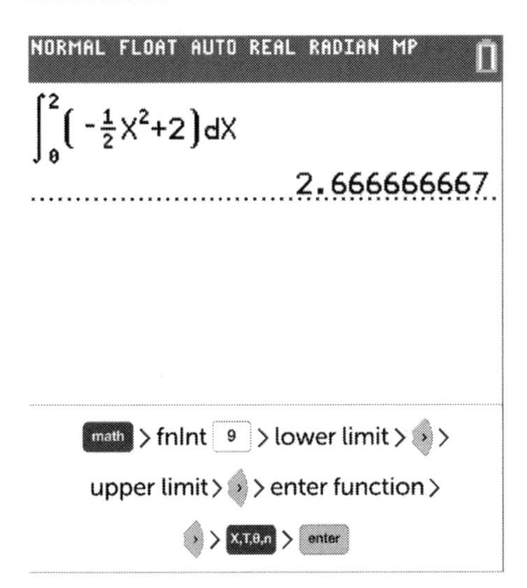

Figure 5-1:
Calculated integral of the function from X=0 to X=2.

5.2 Integral in GRAPH Menu

Press 2nd > CALC trace to access the CALC menu and select ∫f(x)dx 7. Your calculator shows you the graph of the function, and you can set a lower limit. Use the arrow keys ◂ ▸ to move the cursor to your desired location and press enter to set the limit. Alternatively, you can enter an x-value and press enter. Do the same for the upper limit.

Attention: To calculate the area of a function, that intersects the x-axis, you have to integrate from zero to zero and sum up the integrals. There is also a way to calculate the area all at once. Read the following chapter to learn more about it.

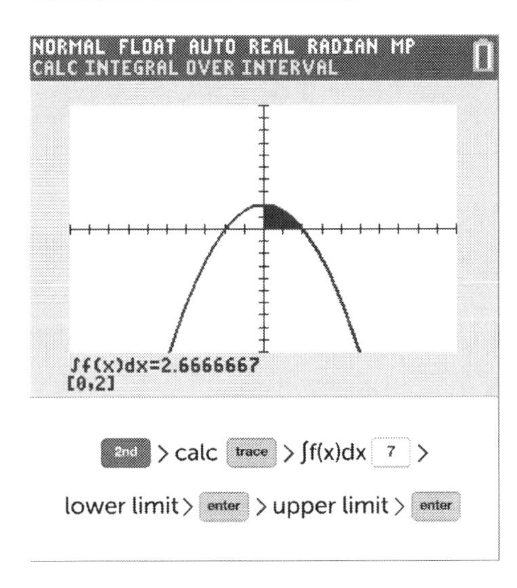

Figure 5-2:
Colored area under the function and x-axis.

5.3 Find Area with Absolute Value

This method will help you to find the area between a function and the x-axis even if the function intersects the x-axis.

Enter your function as Y1 and move the cursor to the line Y2. Press math > to access the NUM menu and select abs 1. Press alpha > F4 trace and choose Y1.

From here, proceed as you would to calculate a normal integral: Press 2nd > CALC trace to access the CALC menu and select ∫f(x)dx 7. Your calculator shows you the graph of the function, and you have to set a lower limit. Use the arrow keys to move the cursor to your desired location and press enter to set the limit. Alternatively, you can enter an x-value and press enter. Do the same for the upper limit.

The zeros of the function are the limits to calculate the area.

Important: Make sure to calculate the integral under the new function (Y2). Use the arrow keys to choose the functions.

It is useful to deactivate the function Y1 (then it won't be drawn). To deactivate, position the cursor over the equals sign and press enter.

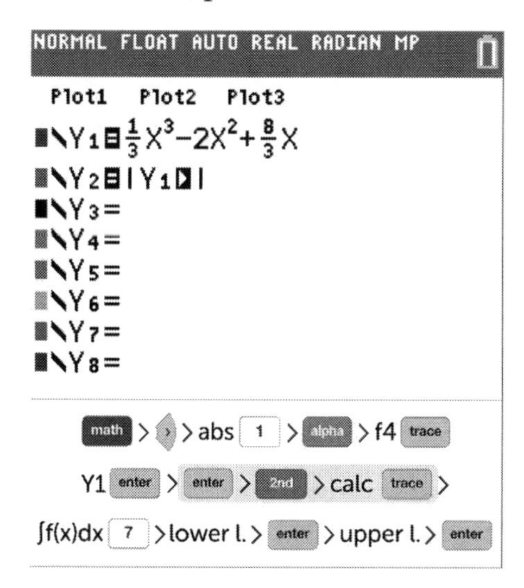

Figure 5-3:

Drawing Y1 with absolute value to calculate the area between the function and x-axis at once.

5.4 Area between two Functions

Enter both functions and press math > to access the NUM menu and select abs 1 . Press alpha > F4 trace and choose Y1, press − and insert Y2. This will subtract both functions, so you can calculate the area between both functions.

From here, proceed as you would calculate a normal integral: Press 2nd > CALC trace to access the CALC menu and select ∫f(x)dx 7 . Your calculator shows you the graph of the function, and you have to set a lower limit. Use the arrow keys to move the cursor to your desired location and press enter to set the limit. Alternatively, you can enter an x-value and press enter. Do the same for the upper limit.

Important: Make sure to calculate the integral under the new function (Y3). Use the arrow keys to choose the functions.

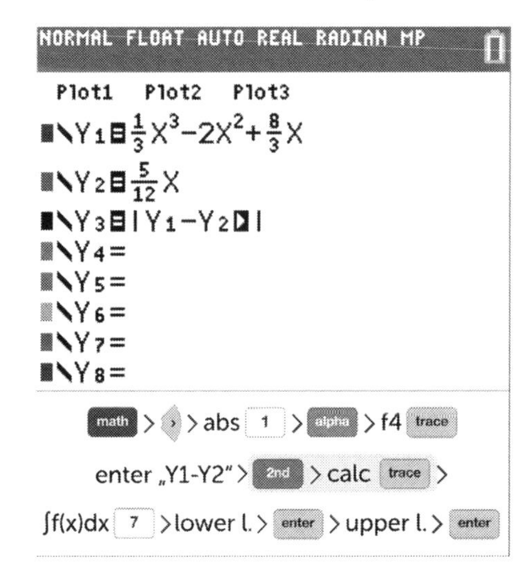

Figure 5-4:

Subtract Y1-Y2 with absolute value to calculate the area between both functions.

5.5 Integral Function

Follow the steps in Figure 5-5. You must insert one fixed limit (lower or upper) and the other limit will be variable. The y-value of the function will be the value of the integral. For instance, if you enter 0 and X, the y-value of the new function will be -8/3 at x=2, which is the same as the integral from 0 to 2.

Xres: This value sets how exact the function will be drawn. "1" is very exact but needs more calculation time.

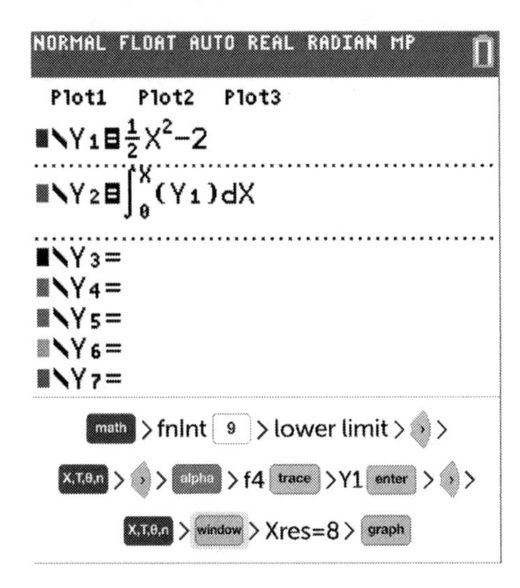

Figure 5-5:
Graphing the integral function of Y1.

6 MATRICES

6.1 Save Matrix

This will be the first step before you can do any further calculations with a matrix. You can choose the letter (A, B, C, etc.) under which you want to store the matrix. Follow the steps in Figure 6-1.

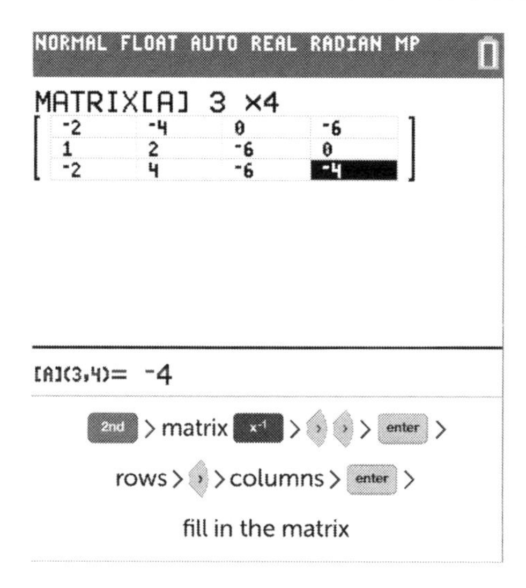

Figure 6-1:
Steps to input and store a matrix.

If you need the matrix only one time, you can enter it with the matrix shortcut menu, located under alpha > F3 zoom. Then highlight the dimensions you want, move the cursor to OK and press enter.

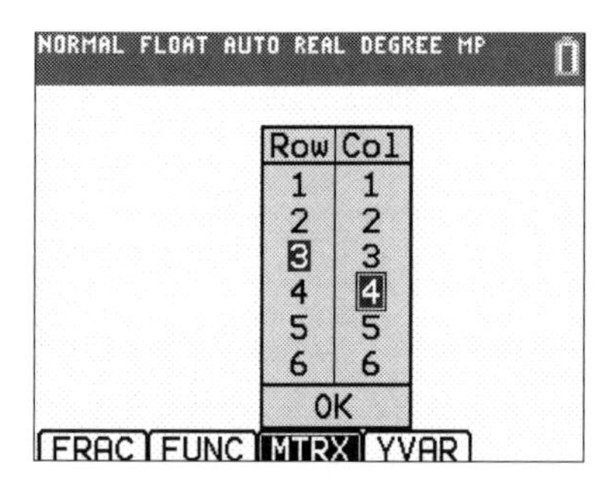

Furthermore, you can store any matrix you've created or calculated. Press sto→ > 2nd > MATRIX x⁻¹ and choose the letter under which the matrix will be stored (A, B, C, ...), then press enter.

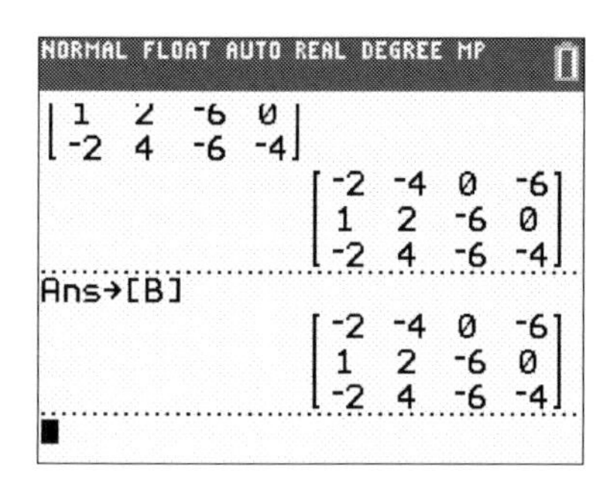

6.2 Delete Matrix

Deleting a matrix is not as easy as you'd think. Therefore, it makes sense to overwrite the existing matrix if the old matrix is no longer needed. It's faster than deleting it.

To edit the matrix, go to 2nd > MATRIX x^{-1} and press the right arrow key ▸ ▸ two times to access the EDIT menu, choose your matrix and press enter.

If you want to delete a matrix, you can go to the MEM menu by following the steps in Figure 6-2.

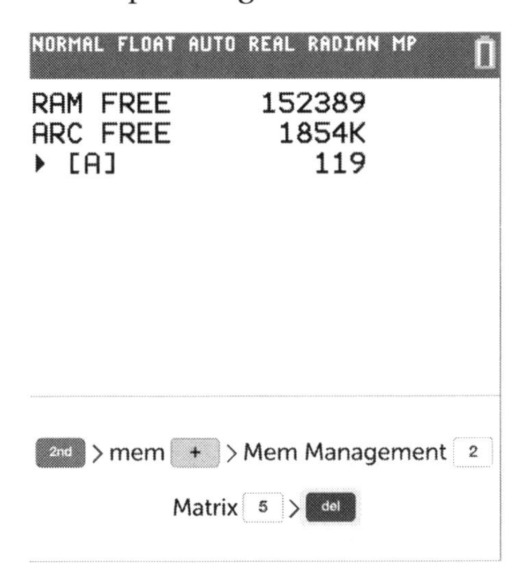

Figure 6-2:
Menu to delete a matrix.

6.3 Put into Row-Echelon Form

Follow the steps in Figure 6-3 to put your matrix into row-echelon form.

This command doesn't solve the matrix. In most cases, you will need to find the reduced row-echelon form of a matrix to solve it, which would be the command "rref(".

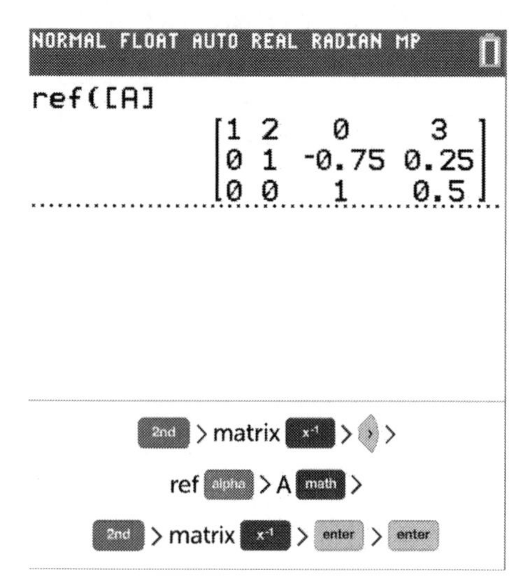

Figure 6-3:

Matrix put into row-echelon form.

6.4 Solve Matrix (Reduced Row-Echelon Form)

Following the steps in Figure 6-4 will find the reduced row-echelon form of a matrix. You can also use it to solve a system of linear equations. The output of the example above shows the solution x=1.75, y=0.625, and z=0.5.

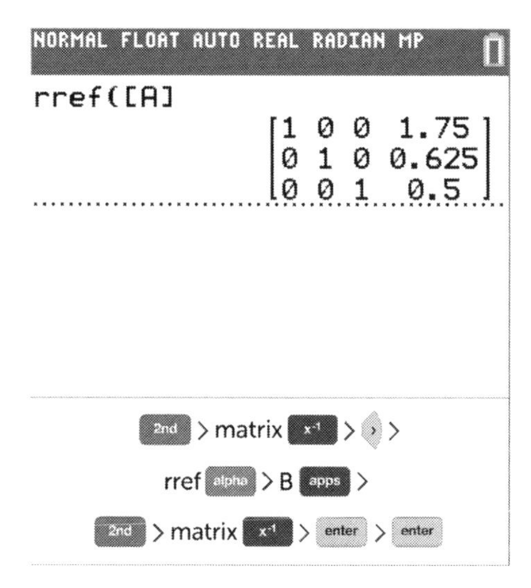

Figure 6-4:
Matrix put into reduced row-echelon form.

Keep in mind: Not all systems of linear equations have unique solutions like the example above.

1. The system has no solution if one diagonal element is equal to zero. Another condition is that the number in the right column of the same row isn't equal to zero.
2. The system has infinitely many solutions if the diagonal element and the number in the right column are equal to zero.

6.5 Transpose a Matrix

Transposing a matrix means turning all the rows of a given matrix into columns and vice versa. The matrix gets flipped along its main diagonal. Follow the steps in Figure 6-5.

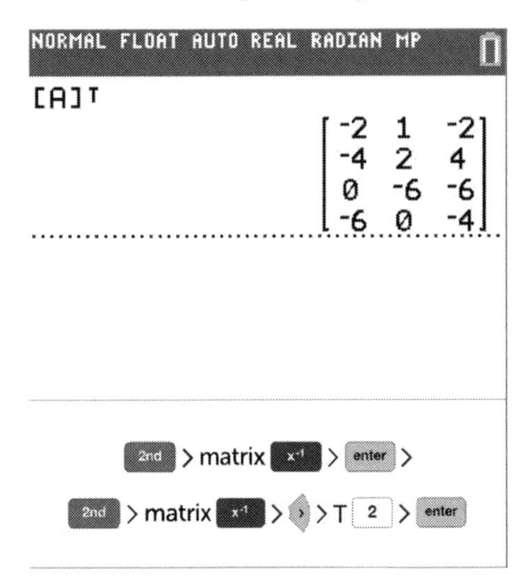

Figure 6-5:

Transposed matrix.

6.6 Identity Matrix

The identity matrix always has as many rows as columns. For example, enter 4 after the "identity" command to get a 4x4 identity matrix. Follow the steps in Figure 6-6.

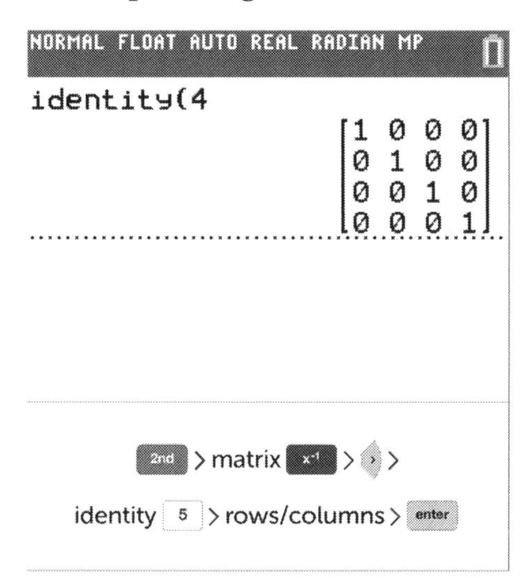

Figure 6-6:
Identity matrix with four rows and columns.

6.7 Inversion of a Matrix

You are only able to calculate the inversion of a matrix with the same number of rows and columns. Follow the steps in Figure 6-7.

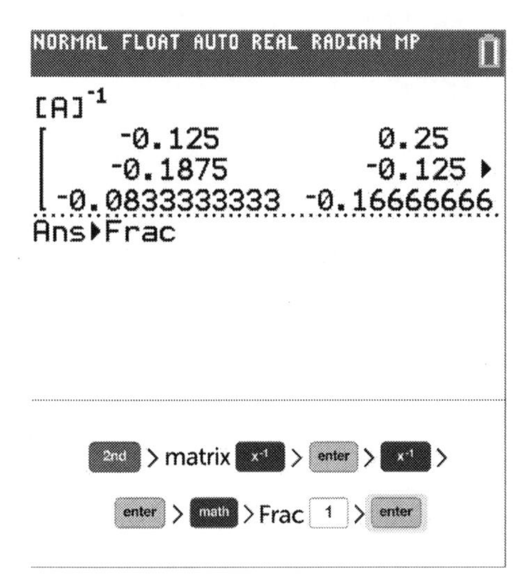

Figure 6-7:

Inversion of the matrix A. The >Frac command will display the answer in fraction form.

6.8 Determinant of a Matrix

You are only able to calculate the determinant of a matrix with the same number of rows and columns (it must be square). Follow the steps in Figure 6-8.

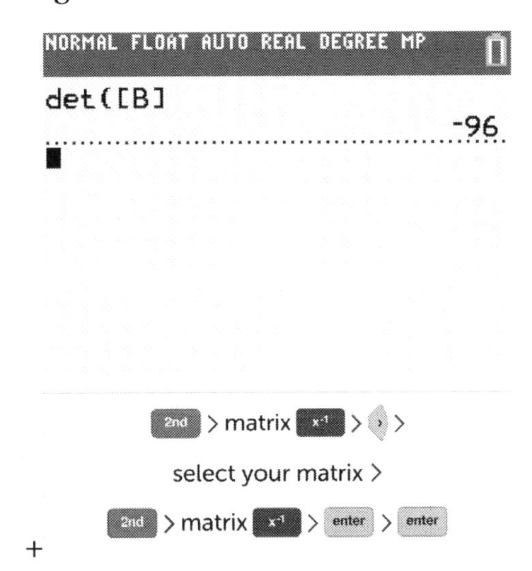

Figure 6-8:

Determinant of matrix A.

6.9 Matrix Arithmetic Operations

- **Scalar Multiplication**: Enter the value of the scalar multiple and then paste the name of the matrix.

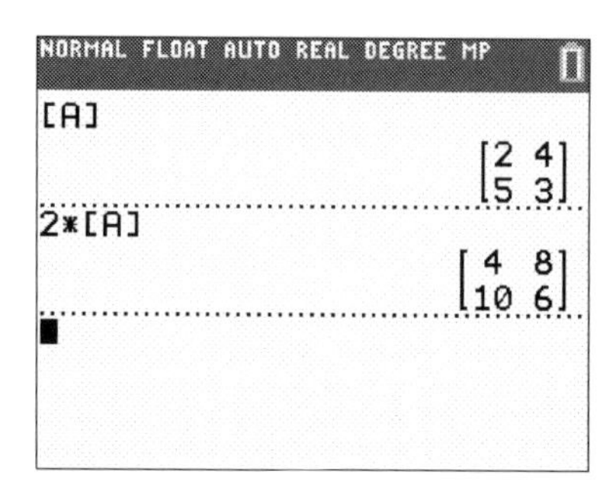

- **Addition & Subtraction**: Note that both matrices must have the same dimensions. Just combine them with + or -.

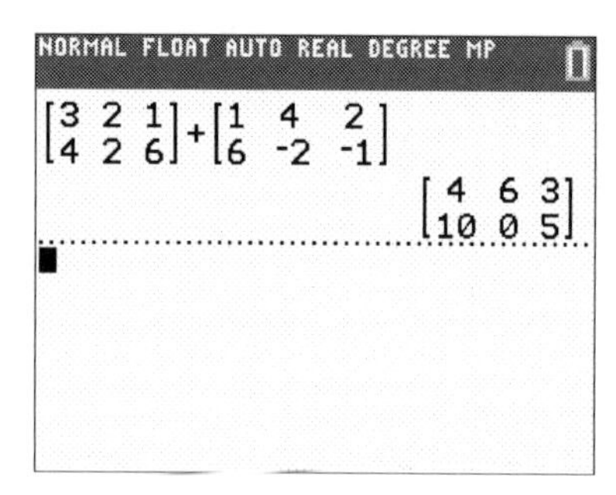

- **Multiplication**: Note that the number of columns in the 1. matrix must equal the number of rows in the 2. matrix. Press * to multiply the matrices.

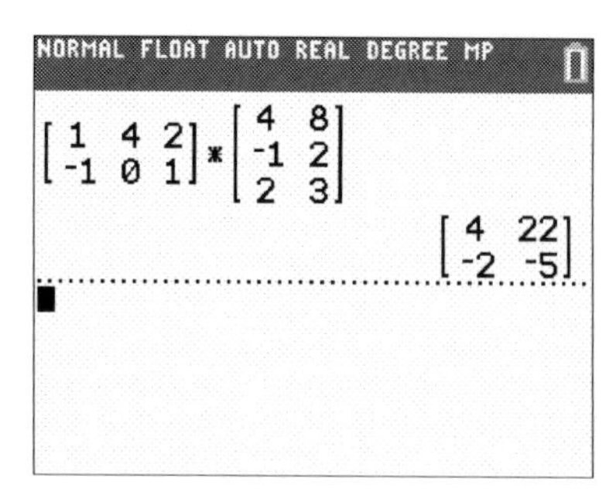

- **Power of a Matrix**: The matrix must be square to find the power, and only positive integers can be used.

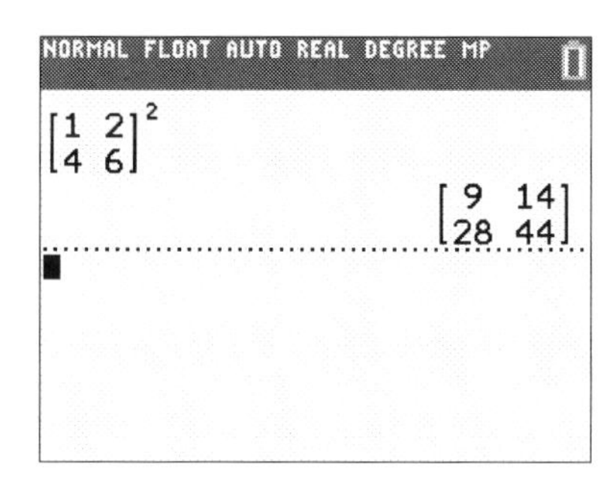

7 STATISTICS & PROBABILITY

7.1 Permutations, Combinations & Factorials

a. Factorials

You often need factorials to solve probability problems. To evaluate factorials, enter the number and press math > PROB ◂ ◂ to access the probability menu. Press 4 to enter the factorial symbol "!" and press enter.

As a quick reminder: 5! = 5*4*3*2*1.

```
5!
                         120
5*4*3*2*1
                         120
```

b. Permutation and Combination

Permutation:

You have a given a set of different items (**n**), in how many ways can you select <u>and order</u> a specific number (**r**) of them?

Example: In how many ways can a president, a treasurer and a secretary be chosen from among 7 candidates? Because each position is different, the order is important. n = 7 candidates, r = 3 positions. It is a permutation and you should use **nPr**.

Combination:

You have given a set of different items (**n**), in how many ways can you select a specific number (**r**) of them?

Example: In how many ways can 3 equivalent positions be divided among 7 candidates? Because all 3 positions are equal, the order is NOT important. n = 7 candidates, r = 3 positions. It is a combination and you should use **nCr**.

Use your calculator to solve the problem: Start on the home menu and enter the number for **n**. After that press math > PROB ◀ ◀ to access the probability menu. Press **nPr** 2 to enter the permutation command or press **nCr** 3 for combinations. Enter the number for **r** and press enter.

$_7P_3$

210

$_7C_3$

35

c. Binomial Theorem

If you are dealing with binomials of high degrees, it can be a pain to calculate them by hand. What's the fifth term in the binomial expansion of $(3x+2)^6$?

Use your calculator, together with this formula: **$(nCr)(a)^{n-r}(b)^r$**. Identifying a and b is quite easy as you should be familiar with the binomial formula **$(a+b)^n$**, so **a=3** and **b=2**. The power of the binomial is **n=6**. For r, you have to keep in mind that it is always one less than the number of the term you want to find, **r=4**.

Now plug the variables into the formula **$(nCr)(a)^{n-r}(b)^r = (6C4)(3x)^{6-4}(2)^4$**. Evaluate the formula in your calculator:

$_6C_4*3^{6-4}*2^4$

2160

The fifth term of the binomial expansion of **$(3x+2)^6$** is **$2160x^5$**.

7.2 Random Numbers

a. Random Decimals

Generating random decimal numbers is quite easy. Press math > PROB to access the probability menu. Press rand 1 to enter the random command and press enter. This will generate a number between 0 and 1. If you need a higher random number, add **10*** before the **rand** command, which generates numbers between 0 and 10.

```
NORMAL FLOAT AUTO REAL DEGREE MP
rand
                    .9435974025
rand
                     .908318861
10*rand
                    1.466878292
10*rand
                    5.147019505
```

b. Random Integers

To generate random integers, press math > PROB to access the probability menu. Select randInt(5 to paste in the randInt command.

First enter the lower limit (smallest number) and confirm with enter, then enter your upper limit (highest number). For example if you set "0" as the lower limit, "0" will be the smallest integer you are able to get.

If you need more than one random integer, you can enter a value for "n" which is the number of elements. It signifies how many random integers will be generated. If one number is fine for you, let it empty. Move the cursor over "Paste" and press enter twice to paste the command to the home screen and execute it.

```
NORMAL FLOAT AUTO REAL DEGREE MP
        randInt
 lower:0
 upper:20
 n:5
 Paste
```

```
NORMAL FLOAT AUTO REAL DEGREE MP
randInt(0,20)
                              8
randInt(0,20)
                             15
randInt(0,20,5)
                  {0 7 20 4 16}
```

c. Random Integers No Repetition

The calculator could generate the same integer twice. To avoid this, use the command **randIntNoRep,** which generates integers with no repetition. Press math > PROB to access the probability menu. Select randIntNoRep(8 to paste in the command. Enter a lower and upper limit and optionally the number of elements "n".

```
randIntNoRep(0,5)
                  {0 2 1 4 3 5}
```

7.3 Lists & Statistical Data

a. Enter and Delete Lists

To enter the STAT list editor, press stat > Edit... 1. To add values to an already existing list, use the arrow keys to place the cursor on the underscore characters located below the name of the list (L1, L2, ...).

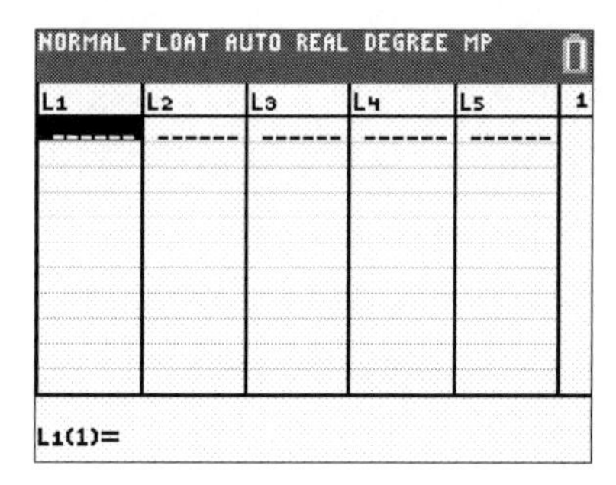

Now enter the values and press enter after each value.

To delete data from a list, use the arrow keys to place the cursor on the name of the list. In the screenshot you see that the first row and column "L1" is highlighted in black.

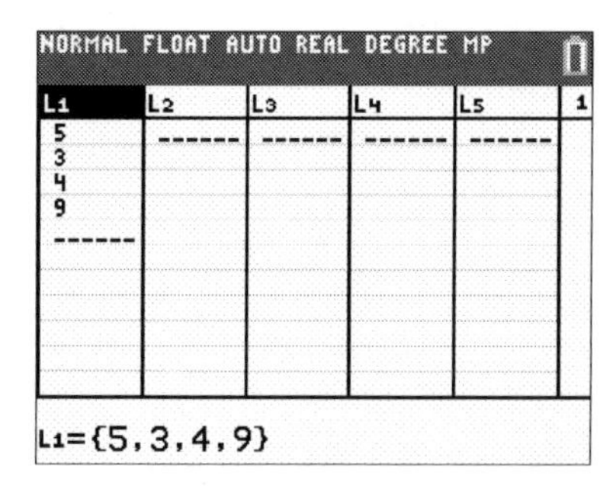

Now press clear and confirm with enter, which deletes all entries in that list. The empty list will remain.

To delete the entire list, including all data in that list, place the cursor on the name of the list and press del. However, I don't recommend using the del button because it is faster to clear all data in a list and refill it instead of deleting it completely and creating a new list afterward.

If you want to delete several lists at once (only the values inside the list, not the names), accessing the MEMORY menu by pressing 2nd > MEM + will save you time. Press **ClrAllLists** 4 to paste the command to the home screen and press enter to clear all lists.

b. Insert a New List

Inside the STAT list editor, you can insert a new list by placing your cursor on the name of an existing list. Press 2nd > INS del to insert a new list. It will appear left of the list highlighted by the cursor. Give your list a name (maximum of five characters) and press enter.

c. Using Formulas

You can use formulas to enter data, for example, to enter the sequence of 5,10,15, ...,100 automatically.

Press stat > Edit 1 to open the list editor. Place your cursor on the name of an existing list, which will be the list to store the new data in. To open the wizard to enter the formula, press 2nd > LIST stat and ▸ to access the OPS menu. Press seq 5 to open the wizard:

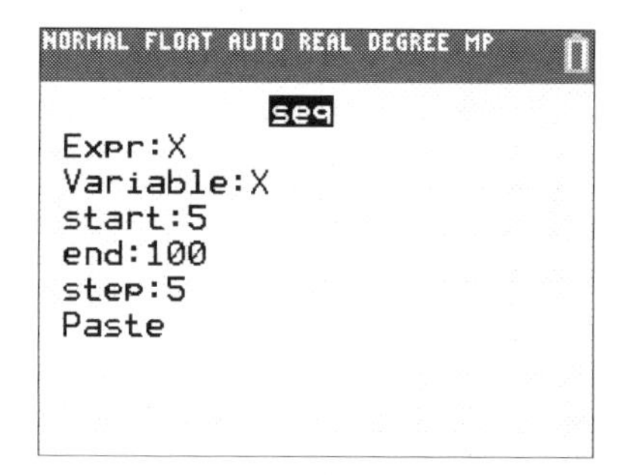

Enter X as the expression and X as the variable. Fill in the start and end value, which I've set to 5 and 100. Step is the increment from one term to the next. Use the arrow keys to navigate to and highlight **Paste**

when you are done, and press enter. Press enter once more to execute the command and to create the list.

d. Sorting Lists

Lists can be sorted in ascending (SortA) and descending (SortD) order.

Enter your list and go to the home screen. Press stat > SortA 2 to sort the list in ascending order or press SortD 3 to sort descending. Enter the list name by pressing 2nd > LIST stat and use the arrow keys to paste in the name of the list you want to order. Finally, press enter.

If you are using the default list names (L1, L2, ...) you can quickly enter them by pressing 2nd > 1 to 6.

7.4 Histogram, Box Plot & Scatter Plot

a. Histogram and Box Plot

A histogram and box plot use data from one variable only (one list). To plot a histogram or box blot, enter your data in the STAT list editor first. Press stat > Edit 1 to access it and enter your data.

To construct the histogram or box plot, a plot must be turned ON. To do so, press 2nd > STAT PLOT y= and press 1 to edit the first plot. Use the arrow keys to highlight ON and press enter.

Highlight the plot type:

- for a histogram
- for a box plot

Afterwards, press to move the cursor to the **Xlist** line where you need to enter the name of your data list. Press 2nd > LIST stat to see all available lists, select the one you want and press enter. Usually, the default value of **1** for the **frequency** is fine. Additionally, you can set a color by using the right and left arrow keys to scroll through all available colors.

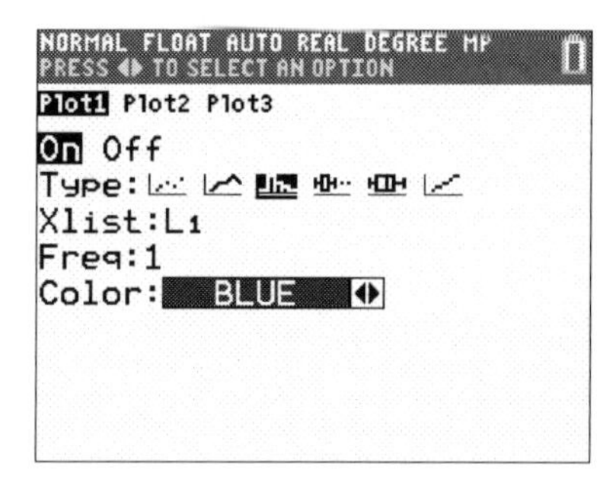

The best way to plot your data is to use the ZoomStat command, which uses an almost perfect window to display a histogram and box plot. Press zoom > 9 to use the zoom command and to plot your data.

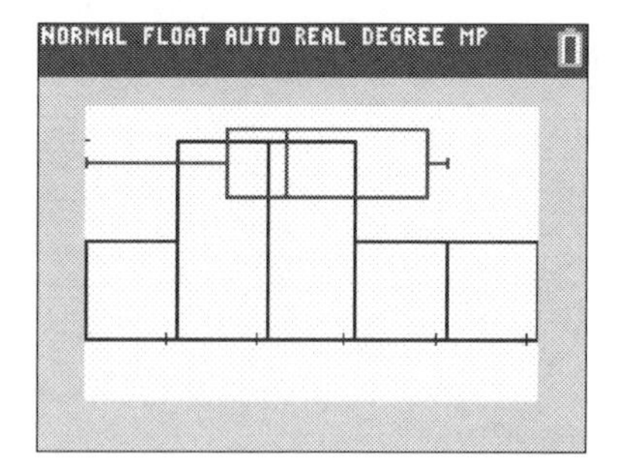

Turn off any functions entered in the y= menu to make sure they won't get graphed. To deselect functions, use the arrow keys to place the cursor on the equals sign and press enter.

b. Adjust the Class Size

The data of the histogram above is grouped into 5 classes by your calculator. However, if you don't like how the data has been grouped, you can change the class size. To do that, go to the window menu and change the value of **Xscl**. A higher value means that it puts more data into one class, which will generate fewer bars.

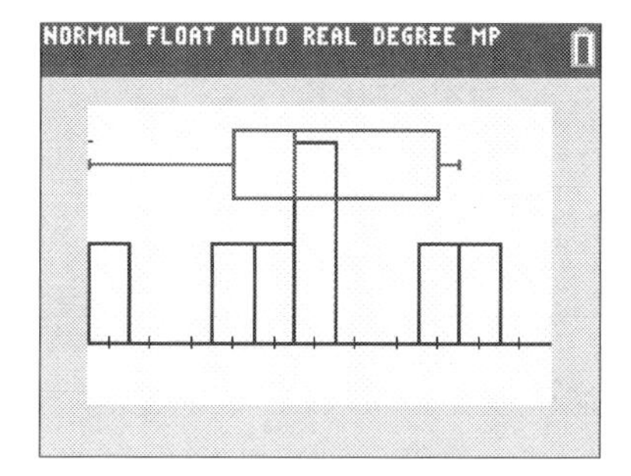

For this example, the value of Xscl is reduced from 4.5 to 2.

c. Two-Variable Data Plots

Two-Variable data plots are plots which use two lists of data. One list will be the x-coordinate and the other list will be used for the y-coordinate. The most common plots are the **scatter plot** and the **xy-line plot**.

Follow the same method described for a histogram and box plot. Make sure you enter two data lists in the STAT list editor first. Press stat > Edit 1 to access it and enter your data.

NORMAL FLOAT AUTO REAL DEGREE MP

L1	L2	L3	L4	L5	2
5	4	------	------	------	
22	40				
15	18				
12	13				
13	13				
16	35				
23	30				
------	------				

L2={4,40,18,13,13,35,30}

To plot the two-variable data, press 2nd > STAT PLOT y= and press the number of the plot you want to edit (1,2,3). Use the arrow keys to highlight ON and press enter.

Highlight the plot type:

- for a scatter plot
- for an xy-line plot

Afterwards, press to move the cursor to the **Xlist** line, where you need to enter the name of your data list. Press 2nd > LIST stat to see all available lists, select the one you want and press enter. Repeat the process for the **Ylist** and choose the mark type.

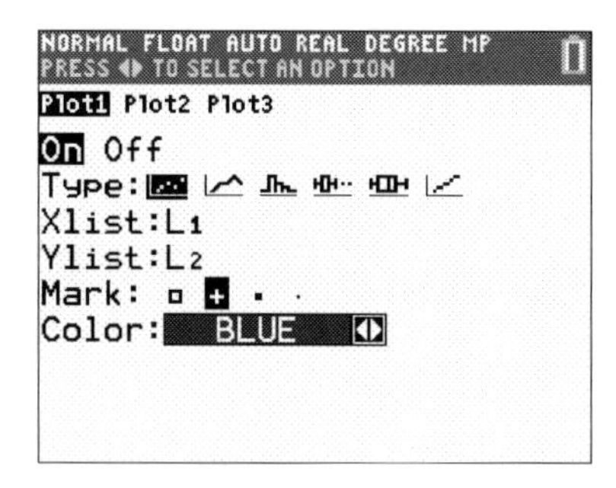

The best way to plot your data is to use the ZoomStat command, which uses an almost perfect window to display a scatter and xy-line plot. Press zoom > 9 to use the zoom command and to plot your data.

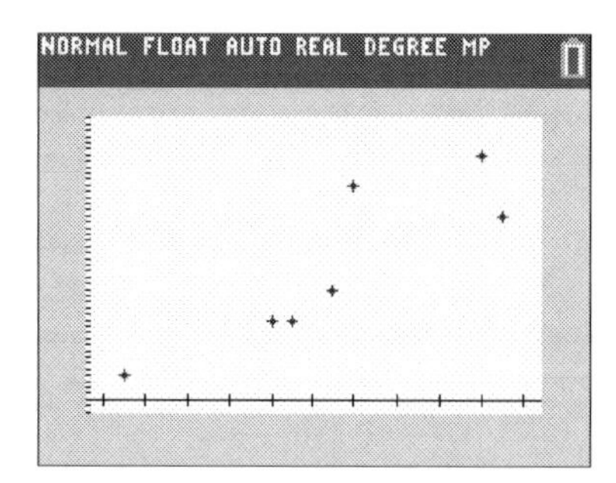

d. Tracing Plots

Any statistical data plot can be traced to see the value of the data. Press trace and use the arrow keys to trace the plot. If you have more than one plot, you can use the up and down arrow keys to choose between the plots.

7.5 Statistical Data Analysis

a. One-Variable Data Analysis

Enter your data in the STAT list editor first. Press stat > Edit 1 to access it and enter your data.

Press stat > [right arrow] to access the STAT CALC menu and choose 1-Var Stats 1. The calculator is now asking for your data list: Press 2nd > LIST stat to see all available lists, select the one you want and press enter.

If necessary, you can provide a FreqList or skip this step (leave it blank). Press [down arrow] to highlight **Calculate** and press enter. The calculator shows you all the statistical data. I'm using the same lists from the previous chapter in case you need an example.

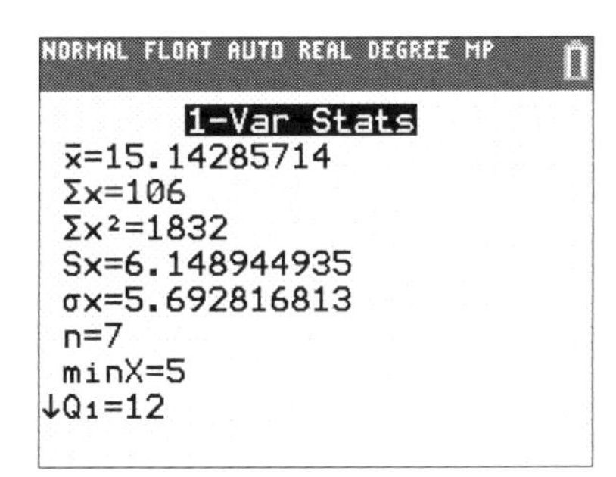

The One-Variable command calculates the following statistical variables:

- $\bar{x}$ is the mean (average) of the elements, as returned by mean(
- Σx is the sum of the elements, as returned by sum(
- Σx^2 is the sum of the squares of the elements
- Sx is the sample standard deviation, as returned by stdDev(
- σx is population standard deviation
- n is the number of elements in the list, as returned by dim(
- minX is the minimum value, as returned by min(
- Q1 is the first quartile

- Med is the median, as returned by median(
- Q3 is the third quartile
- maxX is the maximum value, as returned by max(

b. Two-Variable Data Analysis

Enter two lists of data and press stat > to access the STAT CALC menu and choose 2-Var Stats 2. The calculator is now asking for your **Xlist**: Press 2nd > LIST stat to see all available lists, select the one you want and press enter. Repeat the process for your **Ylist**. If necessary, you can provide a FreqList or skip this step (leave it blank). Press to highlight **Calculate** and press enter.

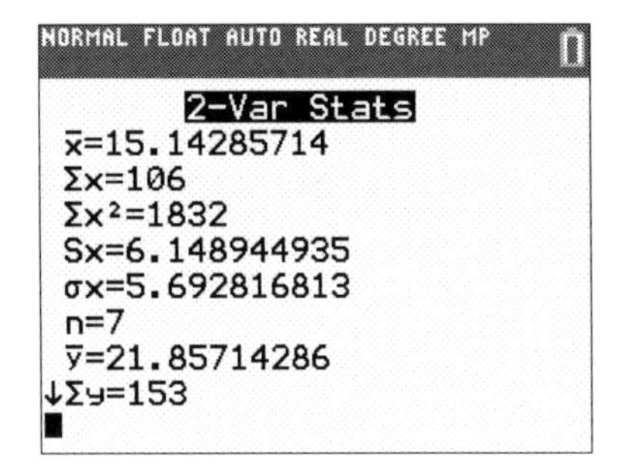

The Two-Variable command calculates the following statistical variables:

- $\bar{x}$ is the mean (average) of the first list
- Σx is the sum of the first list
- Σx^2 is the sum of the squares of the first list
- Sx is the sample standard deviation of the first list
- σx is population standard deviation of the first list
- n is the number of elements in both lists
- y is the mean (average) of the second list
- Σy is the sum of the second list
- Σy^2 is the sum of the squares of the second list
- Sy is the sample standard deviation of the second list
- σy is population standard deviation of the second list

- Σxy is the sum of products of each matching pair of elements in the lists
- minX is the minimum element of the first list
- maxX is the maximum element of the first list
- minY is the minimum element of the second list
- maxY is the maximum element of the second list

c. More Statistics Commands

Some more statistics commands are located inside the LIST MATH menu. However, you get the data for most commands by using the **1-Var Stats** command. If you only need one value, it makes sense to use the commands of the LIST MATH menu.

Press 2nd > LIST stat > ◀ ▶ to access the LIST MATH menu. Choose a command: for example, press variance 8 to paste the command to the home screen. Press 2nd > LIST stat to see all available lists, select the one you want to use and press enter.

```
variance(L1
                 37.80952381
```

7.6 Regression

Let's assume you have four points and want to find the equation of a cubic function.

- f(-5) = 0
- f(-2.5) = 0.5
- f(0) = 0
- f(2.5) = -0.5

The value inside the parentheses is the x-value. At x = -5 the function has a y-value of 0.

To find the equation of the function through regression, enter all x-values in one list and all y-values in another list. Follow the steps in Figure 7-1 for step by step keystroke sequences.

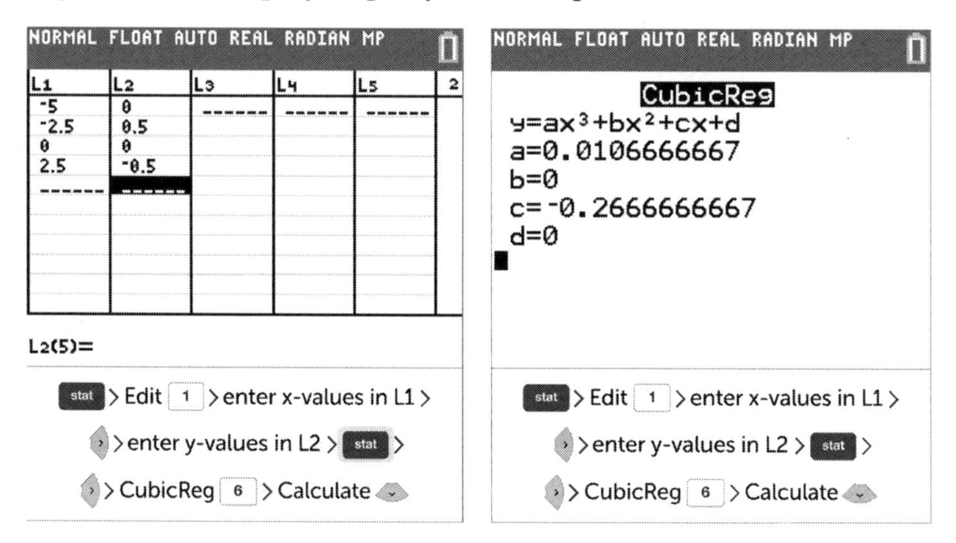

Figure 7-1:

List editor and Cubic Regression menu.

There are different types of regressions, choose...

- **LinReg** for lines
- **QuadReg** for parabolas
- **CubicReg** for functions of degree 3
- **QuartReg** for functions of degree 4
- **LnReg** for logarithmic functions
- **ExpReg** for exponential functions
- **SinReg** for trigonometric functions

You can store the equation as a new function. Just enter for example "Y1" (alpha > F4 trace) in the field "Store RegEQ".

Made in the USA
Middletown, DE
21 August 2019